P9-CKA-513

THE THIRD LEAF

The Third Leaf

THE BIRTH OF A VINEYARD

BETTY WILLIAMS

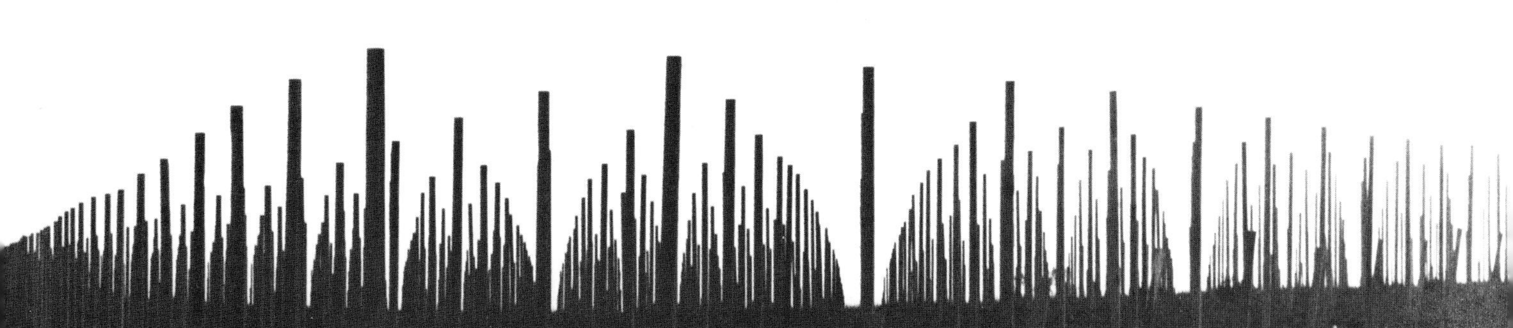

John Daniel, Publisher
SANTA BARBARA
1986

PRINTED IN THE UNITED STATES OF AMERICA

LIBRARY OF CONGRESS CATALOGING-IN-PUBLICATION DATA
Williams, Betty, 1918-
THE THIRD LEAF.
1. Viticulture—California—Santa Ynez Region—Pictorial works.
2. Vineyards—California—Santa Ynez Region—Pictorial works.
3. Santa Ynez Region (Calif.)—Description—Views.
4. Williams, Betty, 1918- W. Title.
SB387.76.C2W55 1986 634'.8'0979491 85-24715
ISBN 0-936784-05-9 (pbk.)

Published by
JOHN DANIEL, PUBLISHER
Post Office Box 21922
Santa Barbara, California 93121

NOTE

You plant the rooting in late spring. It produces the first leaf. The following winter the plant is cut back almost to the ground. The new growth is the second leaf. Training begins with this growth. The following spring produces the third leaf and the first grapes.

PREFACE

STARTING FROM SCRATCH, bare dirt and ignorance, gave the process of planting our first vineyard a dark aura, murky with uncertainty and contradiction. Reading and listening did not dispell the confusion. One at first concluded that everyone knew how except us. Later, it seemed no one knew. The whys and the hows lived in a marvelous world of creativity—scientific, mystical and practical. The frame is large and at times a bit sloppy. Killing a whip of *vitis vinifera* is not easy. But first things first. It does have to get into the earth.

This is the story of how some thirty thousand examples of that remarkable grape went into the ground on the mesa above my pastures. Rereading creates in me a glow of embarrassment, a recognition of my cheek, innocence and ignorance. Nevertheless, I've come to see the idiocy and charm of this pursuit and I've lived long enough to know that some of life's greatest charms are its idiocies—making a vineyard has to be one of those. It only leads to the further hopes and joys of the final end, the making of the wine itself.

A sentimental penchant for documentation led me to a diary and for photography, a natural inclination to reduce any process to images for posterity. Here is the story of our planting, somewhat incomplete and ridiculously human. Feel free to learn from our experiences, but under no circumstances listen to any advice any one of us may be constrained to give.

The wine community of our valley is full of friends and cooperative spirits. Helping us get started seemed to be an easy part of their lives. Michael Benedict has been a fast friend without whom much might never have been done. To all those with whom I spoke and questioned and to the men and women who labored in the vineyard and especially to Bret Davenport, my son-in-law and colleague, on whom the heavy work was laid, my deepest thanks.

INTRODUCTION

The journal of a vineyard in the making

October 4, 1982

DRIVING THE BIG BACKHOE, Tom made steps in the ground going the full ten or twelve feet. We walked down. Michael said, "Get out your pocketknife, Betty," just as if he thought I always carried one. Fortuitously, I'd just bought one for the car. We chipped away in our narrow vertical cave, going from top to bottom, testing the density, appearance and texture. A few feet of top soil, a few feet of clay, some shale and some sand, all of these varying for each site.

We looked, too, for roots, tiny signs of growth through the compacted clay. There were some in most areas despite the fact that I had not had deep rooting crops grown above. We found one bad area, actually two holes about thirty feet apart, where there was hardpan about five feet down. This was a bad sign, but Michael said if it's not widespread, which we can find out next week with an auger, then it's ok. The best areas are two wide strips on the south face of the big hills in the back which will run east and west. We had checked the profile of the land for use in wine grapes.

October 14, 1982

AN EXCITING DAY with mixed returns. I started out to a vineyard again to take early shadow pictures. I climbed around here and there. I knelt down low and looked up into the sun. I ended the roll and wound it to the finish. I opened it and there it was at the start, jammed. Not a picture in the whole roll and far too late to be able to take them over.

Michael arrived at Buttonwood later in the day. We went up in the back for two hours planning the rows and roads. We stayed until after dark, tramping up and down, eyeing the swale, the slope, the complex curves, learning the pattern of the ground and planning the best use of it. Occasionally I looked around at the evening light, the sunset, and felt terribly small and a little afraid and had to remind myself I was only a half mile from home. And I thought of all those rows and rows of grapes I would be nurturing. Later, I told Michael of my hakea tree and my cypress, brought up from milk cartons, now four and fifteen feet tall. He said, "That's all it takes. Your grapes will do well."

October 17, 1982

TOM CAME IN THE following morning and started ripping. I decided on the entire area, all the fifty acres that can be done, leaving out only the steep peripheries and the little glades and glens or areas with trees. I should be able to get a good crop before we plant grapes and that should pay for the work. My crop problems before have largely been lack of sufficient ground preparation. Anyway, I expect a big water year and if it doesn't wash the seed or come at harvest then I should win this one.

October 30, 1982

In the meantime we have pursued the wine business. Bret is willing to help and seems very interested, said he'd get books on viticulture and wine making. We have talked more to Michael Benedict. Frank, Bret and I have talked to Allen Russell at the Firestone winery, and to the winemaker at the Ballard Canyon Winery, and to a woman at the Carey Vineyards. They are all certainly different from Michael, who disciplines himself in a very European old world manner. His beautiful collection of kegs, barrels and vats makes a lovely sight. The others all have some huge, if also handsome, steel vats that are entered from the base. Michael's are all buckets heaved into the open top. Some of the others are even out-of-doors. He has clearly the best arrangement, even though it is austere. I think that we shall not be quite so rigid although it is beautiful and makes fine wine.

Meanwhile, all our subsoiling has been done, back and forth about three feet. I will try now to get Brierly to disc and plant barley for all but the first 12 acres we are to plant. And that, now, means ordering the plants.

November 10, 1982

Winter time! It froze last night. The early evening was a brilliant cold, still night. I called Seeb to come back up after dinner to help me take my porch plants into the greenhouse. I have no heat there but it stays several degrees warmer, enough to save the plants. I shall know later.

Jerry Simpson said I'd need overhead sprinklers for frost control and deer fencing, both contrary to Michael and Allen. He also said I should get at least four tons to the acre or I could not be commercial. He was mildly scornful of dry farming. "I know irrigation blows up the grapes but you have to have it. We do it and I think we put out respectable wines." They do. And they are getting better. But the vines are maturing too and that helps. This is certainly a mixed bag, from Michael's austerity to Jerry's freewheeling.

January 24, 1983

TODAY I'VE been making calls about buying rootings. Saturday we talked about cuttings and decided I could pick them up from Michael and from Brander as canes and bring them home and make the cuttings. From there we'll bury them in the prescribed fashion. Soon we'll be laying out the land with all those stakes and end posts until it looks like a military cemetery. Then we'll take all those plants I just ordered and which probably will also have been at least partially buried in the ground and put them properly in the ground by each stake. Then the care begins.

February 8, 1983

WE STARTED doing some of the cutting of canes from the Brander vineyard. It has been delayed by more and more rain. It gets so sopping wet in the vineyard that you can hardly move from vine to vine. We are stashing them in the boxes that were made for trash bins for Seyburn's wedding, a wonderful idea because it was very hard to make them stay upright. As it was, half of the first three thousand we have are not upside down as they should have been.

May 20, 1983

MICHAEL WENT over my nursery. He approved, said my system was fine including the heretical method of using a trencher to make rows for the cuttings. He said now when it starts up top it has to go like a fire drill. He was pleased with the irrigation people's effort to save me several thousands because of their price war on irrigation equipment. So I feel good about the decisions I've been making that sometimes make me antsy and lonely.

June 8, 1983

TIME TO BEGIN. Yesterday was busy all day and exciting. Kate had come up unexpectedly so she got in on the first things. Michael arrived with two rasty Mexicans to do root pruning. One cuts the roots in the shape of a hand with three-inch fingers, leaving the strongest branch at the top, cut to just above the second node. We all learned.

It is getting to be like the fire drill I am to prepare for. I had a farmer doing the last of the discing under pressure from me and the pressure of dovetailing. The surveyors were due this morning and the farmer wasn't done. They couldn't be kept waiting because the irrigation people were going to start Monday. And Tom needed to get with the roads at the same time. First things first. It has been an acute and remarkable performance getting everything going in order. So far—so very good. I'm feeling mighty pleased. We ARE getting in a vineyard.

Now I need several feet of nine different colors of synthetic cloth to be tied to a magical length of cable which will tell us exactly where within a fifty-foot strip to put five-foot and ten-foot pegs into the ground.

Grids, transits, the lovely curvature of the earth and all its sundry miniscule micro curves plus lumps and bumps. Tags and numbers and all things to confound, bewilder and bedazzle. When it's all over, we shall have a vineyard ready for all those cares and kindnesses.

The excitement of setting out the gridding is immense. To go to the top of the hill and look over the twelve acres to be planted, to let the eye wander over as much as one can see and dream of the rows and rows of grape vines—that is heady indeed. I now know that in the grape parlance 'heady' means high in alcohol.

June 24, 1983

THE GRIDDING goes on and is becoming a pleasure, especially being able to look over the hill, seeing all the white stakes in those gorgeous patterns. The planting crews are growing as I recruit students from UCSB. Warren is at work with Bret and the gridders. He's keeping up his gophering but there seems to be no help for moles with their omnivorous habits.

June 25, 1983

TODAY RAMIRO has a small team to rake about an acre that is heavy in old roots. The surfaces have been too bumpy and rough. It will ease the planting.

July 7, 1983

I MUST GO and get more gopher traps. The earwigs are being poisoned with a five-percent Sevin pellet which we are putting down at about three pellets per plant. More, apparently, is redundant. Watering has been done on the whole first block but it has to be repeated. It didn't work properly. Now it's being put on the north thirty-two rows of the east bank where they're still planting.

July 10, 1983

THE PLANTING for the nursery is almost completed for the year with a few hundred extra for ones that gophers will get and edges we've forgotten.

July 13, 1983

WE HAVE HAD numerous crises with the irrigation. The pressure is far too low. A booster pump and a twenty-percent increase in the cost are in the offing.

The early mornings are simply gorgeous—but oh so still—and it means the rest of the day will be scorching hot.

July 16, 1983

BRET HAS JUST left. We have gone over the final things that need to be done. Regarding irrigation, we had the heights and drops checked by the surveyor and Michael with his calculator tells us that if they run a road up there with a three-inch line we can get sixteen pounds static pressure, enough to water the tops of the rows by the simple process of engaging the same hoses and emitters that are coming from the bottom. So now our last twelve plants will be watered. We will not have to put in a pump—not this year.

Daily care of the vineyard means watching the plants, checking for damaged leaves from insect pests, and gophers, which are being trapped by the dozen. In the nursery the moles are a major menace. They run into a row of little cuttings, run along them and move away after about eight feet, leaving eight feet of cuttings without earth packed around their newly formed roots. Having discovered this, we are walking the rows and checking the droopy leaves for a need to repack the ground.

August 5, 1983

WARREN TELLS ME that the irrigation works well if you keep the screens clean and you only run about three lines at a time. That's fine for this year but in a generation we don't want that foolishness. It's only the middle rows that go all the way to the top. We know now we can overcome it but we haven't convinced the engineers yet because they think in pounds of pressure and friction loss, regulators and compensating emitters, when in this case a consideration of gravity is vital.

August 12, 1983

THIS MORNING the water engineers are coming to pit their theories against the force of gravity. Today we set some of the pasture irrigation pipe and their regulators, etc., and find out if water will run downhill.

August 22, 1983

I HAD TO TELL Michael his planting caveat did not work, that we had to put water into the hole. He had told me if I did that the clay would become concrete around the roots. I said I can't get the water to them at all if I don't—so I did. They are in great shape now.

Jim Dana says the soil analysis shows enough nitrogen. What you put in now will only go down and will not be available in the spring. They are about a foot shorter than ones planted at other vineyards in March but he insists they have plenty.

September 4, 1983

THE MAIN EVENT is that by persistence and insistence we have finally got the irrigation problem solved well and at a reasonable cost. But it took a lot of conferring and a last suggestion from Tom Petersen. The engineers still fight the idea that I won't have to have a pump but we have compromised in a sense. The pump will be an inexpensive one, a few hundred as against a few thousand, that will work off the tractor.

Meanwhile, we haven't enough water at the top. I wonder about borrowing the water truck again for some remedial watering until we can get the stuff in. The water level at the valley reservoir is down and so is our pressure. The top vines are suffering again.

September 13, 1983

WARREN AND I marked out the route of the sub-main that will feed the vines from the top of the hill. The basic idea was to make nice contours for a beautiful vineyard. Naturally some of the vines didn't quite fit and there will be new ones to put in. I walked, dragging a stake behind me. Warren followed with a bag of lime. It was a very silly looking scene. I hope no one saw.

September 22, 1983

THE LAST OF the irrigation was completed yesterday and it all worked perfectly. We have only one very bad leak on the brand new valve they put in for the use of a temporary pump, when and if that comes. It must be leaking several gallons an hour and Warren says it's a major problem to fix—wrapping calking in the opposite direction to the turning, thereby causing it to lump. Complicated world.

In the later afternoon, after a while at the desk and a discussion with Warren, Ramiro and Bret about the future employee plans, Michael arrived and we went up top checking out those ten thousand chickadees. We talked about fertilizer. We checked out the final irrigation installation. We talked about roads, about shifting section sizes, types of grapes and amounts of them. They thought of taking out part of a major road and putting in more sauvignon blanc, about a new area for more semillon than we have space for, about the fact that there's not quite enough room for all the red we want.

September 26, 1983

It was a shattering blow when a tour guide at Zaca Mesa told us $2.5 million for the first 20,000-case section of their winery. We hope she didn't really know. This is hardly part of our dream.

October 23, 1983

I doubt that this vineyard is ever going to be really smooth. It should have been easy to disc and seed, but it hasn't been. We had a breakdown on the seeder and then the absolute necessity to take the tractor off the hill and go into the pastures for mowing and irrigation. The tomato harvest is still in the way, too. I can't get any of the farmers to come over and disc on the big part.

October 30, 1983

Bret and I went yesterday to visit the Chamisal Winery. It is small, built with thought for expansion—up to 2,000 cases from a few hundred in three years. They have fifty-two acres in chardonnay and sell much of it. We liked the simplicity of approach, minimal space, and tight use of it.

November 19, 1983

My trip to Bakersfield was wonderful. I flew over to see the plant where the grape stakes are preserved. I couldn't see just ordering them over the phone. Toby said, "You tell them if it isn't right the shipment goes back." But if that happened it would be weeks before I'd have my stakes—so I didn't. The owner met me at the airport, drove me to his place and showed me everything. There wasn't much to see but it was impressively orderly. The stakes were clearly what he had described. We discussed how you get them to the rows, how you use the fork lifts. I had to like the man because he uses the same kind of notebook and same system for his daily chores and phone calls that I do.

January 15, Sunday, 1984

VINEYARD WORK went on. We finished the weeding and two days later all the pruning. My timing on the weeding and getting in a crew of nine was perfect. Armando is a great recruiter. I raised the pay so I'm not getting the reluctant dregs. Jeff handled the pruning by bringing his own crew. Now the big question is the use of the herbicide, Round Up, because we only weeded right around the plants. Grasses are growing and will be well onto six feet before too long, especially on the vine rows. The middles we disced in order to plant the cover crop so there's not as much there. We have word that you can spray the Round Up all over the dormant vines. Word that it will kill one-year-old plants. Word that you can cover the plants with plastic pipe during spraying, taking two or three men instead of one to spray.

January 29, 1984

THE STAKING IS going very so-so in the vineyard. They can't seem to get them straight. I said, "That's not good enough—get the surveyor to come back." Jeff said, "No, impossible." Now they've put in a hundred crooked stakes. I wonder what happens next.

February 25, 1984

IT'S COLD AND WINDY. And too, too dry. It's drying up the vineyard and slowing down the staking. The irrigation is in disrepair. I hate to work on it now. It will be less expensive when we eventually do it and we will be doing a lot of it ourselves. The staking is going fairly well and we will plant with shovels as Jeff suggested. It seems it's done all over and works well. The only problem might be lack of moisture and no moisture is predicted.

April 28, 1984

IT'S STILL COLD and the wind is blowing. It froze night before last and caused damage among the wine growers. I had a number of plants that were bitten but they will just get a new set of shoots to train a little later on. I don't think we were into the area where it froze with the tying. Irrigation work continues in the hopes that we will soon be able to get water on the plants easily. I looked at what seemed a dry area and the plants had all their little antennae sticking up with no signs of stress.

In spite of the weather, the vineyard is stunning. It's an awesome thing to look at considering where we started from. It's hard to believe all that's going in. The new part is getting all its stakes first and planting later. It will certainly get the plants in the right place but the real test will be how well the plants grow through the summer and winter to come.

July 24, 1984

BIG BUCK RABBITS have been killed off eating newly planted merlot. Some of it has been protected by milk cartons. The damage is decreasing, but they do make leggy, weak plants.

November 17, 1984

THE RANCH IS ALIVE, a living thing. Whenever someone leaves, it returns to me with new burdens and decisions. I have to go through files, records and calendars to keep it moving. It makes me think of irritations and prejudice. I have a prejudice that people are supposed to know what they're doing and should do it well, and if they don't, I get irritated and say something destructive or look unpleasant. So I've decided that the whole purpose is the process and that disasters are only things to be avoided if one can.

January 19, 1985

I AM FOREVER intrigued by the vineyard. I walk in it almost every day. I love the organization of the stakes in those fascinating rows that move with you as you move. The work is being done methodically and well. I've a lot of this and that over the stakes. We lost nearly fifteen percent in the last years. I've been told I was wrong about the size and kind of stake. Now I look at it and think, "This is a really good-looking vineyard, and it's the slightly bigger stakes and end-posts that are doing it!" Then I was told this was soft wood and not good, etc. And now I'm being told, "Who knows what's best! Maybe the softer wood is better because you don't have to use a screwdriver to get a staple out. You can put in a nail." Which is exactly what we're planning to do. I hassled over the price and losses and came to the conclusion I would prefer to deal with the man I knew, that it came out about the same, cost one way, loss the other.

March 1, 1985

WINE—WHAT A NUISANCE and what a glory. Do I really want to do this? I stood on the top of the hill asking that question dozens of times before I took the plunge and I'm still asking it.

The new herbicides are going in the vineyard.

Four kinds of grapes are to go in before long. They're all in the round pen, waiting anxiously, and I sit here biting my tongue, blowing my nose, honking and coughing, sweating as I keep warm in order not to be chilled. Nothing gets done. Maybe because it's Friday. It may seem amusing but I find it very dreary. Besides, I'm being forced into chemicals I find obnoxious.

March 16, 1985

MANY THINGS HAVE happened in the vineyard. We are spraying with the various weed killers and a pre-emergent. We've mowed and it looks glorious. I'm thinking that I'll look into having the tying done with twine and stop this mess of little nasty plastic things all over the vineyard. Will probably run into massive resistance. The new spray rig is Bret's baby. He and Warren fixed a small trailer for two men to put staples in at once. I hope they work better than the last ridiculous ag ties that shriveled up in the sun and frost. These are supposed to be impervious, especially treated. And fascinating little gadgets they are.

The gridding has all gone in in the cabernet franc area and most of the staking. Next week that finishes and they go to the cabernet sauvignon. The main trench for the irrigation in those fields is in too.

April 4, 1985

I WALKED UP to the top at the south end and looked back over it all. The vineyard looks stunning. You can see all of it except the part we will be putting in later into semillon and the pond, which is down too low. It has been beautifully done. The men are a wonderful crew. They have made straight lines with the stakes and plants. We have lovely curves in the roads. We have learned to put in our own irrigation. Last year was the first. This time it is going in much more easily.

May 23, 1985

THIS IS THE terribly exciting moment when the vineyard is *in*! And we're not faced with any great problems—the finances, the health or ill health of plants—just the momentary success of having done the first steps well. It's heady and there isn't even any alcohol yet. All during a party Seyburn gave Bret and I took people on tour, enjoying the whole exciting adventure over and over. It was fitting. It was euphoric. It was fun. For a day we could indulge our triumph, and we did.

THE THIRD LEAF

Whether or not to put in the vineyard, to consider a winery was a time for walking all over, up and down the rolling hills on the mesa above Alamo Pintado Road near Solvang in Santa Barbara County, California. After that came the many hours of land preparation done with big bulldozers and tractors, the tines turning the ground into heavy clods, later the discs smoothing it out. The job was difficult because of the 'trash' that lay on the surface, the dry stems and roots of generations of grains that clung to the soil. When we were able to get in to the hills after months of rain the tractors mired, and bulldozers bogged behind them. Later in desperation we designated those areas for a pond and for special drainage.

Roads were planned from the tops of hills. From there we laid out sections for each type of grape, the roads that were for transportation alone, those that would have to be wider to accommodate turning tractors. Though these pictures look as though immense amounts of soil are being moved, we only changed the contour in one place and that was to prevent having a turning area too steep, endangering tractors and their drivers.

As soon as their fields were dry enough to walk in, I was given permission by two of my friends to take cuttings in their vineyards for our nursery. We went in as a family along with our barn people and for several days worked to get thousands of cuttings for use the following year. These long whips are cut into fourteen-inch sticks, a diagonal cut at the upper end, a horizontal cut at the base, clear signs of which way is up. They need to have four or five nodes, the top cut being made just above the last node. When planted only the top four inches stay above ground. Within weeks one of the two top nodes will swell and burst open, in time forming a bush which will not bear any blossoms or fruit. The following year when it has lost its leaves it is taken from the ground and cut back to a single stalk about fifteen inches long. At the base is a ball of roots. These are reduced to three or four short stems known as a hand, and the top to one small crook like a finger with at least two nodes. Known as rootings when they are planted, the same top four inches remain above ground, just as it is with the cuttings. Cuttings too may be started in the vineyard. We have done replacements that way completely successfully. It's best done within hours of clipping.

FRIENDS TOO recommended the commercial outlets from which I could order reliable rootings for the first vineyard plantings. We had bundled the cuttings in packages of fifty. The rootings came in bundles of twenty-five. They were laid on a sawdust surface in the round pen, a roofed but open building where we normally trained young horses, a layer of sawdust between each layer of rootings. They were destined to stay there much longer than intended, for the weather did not allow us to plant until June. But time has shown us that they were not harmed. We watered the mounds a little every day. Few broke out into growth. Our loss that year was five to six percent. It was not until our third planting that we had a spectacular take. That year we planted over ten thousand plants and lost less than a dozen. But we can't really say why.

Our plans were to have rows ten feet apart and plants five feet apart. Surveyors were asked to do a fifty-foot horizontal grid. Then our gridding started in the workshop at the barn where Michael and Bret put together a fifty-odd-foot length of cable. Every five feet this cable was marked in varying spacings approximately an inch apart with nine different colored ribbons crimped onto the cable with a marine device using a soft copper. The process can be seen in a sequence in the pictures. In the vineyard one end of the cable was placed at a fifty-foot marker and stretched on the ground to the next. Since the terrain was hilly the distance had to be a little longer than fifty feet. The cable took care of this. When the second fifty-foot stake struck at the blue ribbon, for example, each worker was asked to put a marker at that ribbon. Think of the ground as the hypotenuse of a triangle made by a grid hanging horizontally in the air. You want the plants to be exactly five feet apart as they are in the air. They won't be if you follow the ground. The cable allows the correct reduction of the spacing. In the subsequent plantings the cable was not used as much because the crew had worked together twice and become very expert at a kind of eye-balling that was their own. Nevertheless, it's a remarkable device to start things out each time and saves the cost of more minute surveying.

TWO MEN and two women are seen here placing the markers at the five-foot points indicated by the cable. When this was finished the planting was begun. From the top of the hill I could see everything in process including some of the main line work on the irrigation.

That first year we planted late in very moist soil and rough terrain. It had been almost impossible to smooth out the surface. We used a two-man motorized auger which drilled six-inch holes about eighteen inches deep. The plant was held in the hole while water and dirt was poured around it. It worked all right but we took a simpler route thereafter—just sticking the shovel into the ground, holding the dirt aside, putting the rooting in and stamping it together. Both ways worked equally well. The first simply took much longer.

Not terribly dramatic it was yet exciting to see the hundreds of plants in the ground and then to see the lines of irrigation set down beside them. The water had been brought up from the connection to the Santa Ynez River Water Conservation District on Alamo Pintado Road. It went through a pasture and turned to go directly east and straight up the hill. At the top it turned south through the vineyard meeting the east-west line of the survey meridian, making a line to service the south portion of the vineyard. Had we been thinking more clearly we would have taken that main line to the meridian by going over the big hill which divided the merlot from the sauvignon blanc. In the end we had to do it. The result was the same—not too much extra cost but unnecessary.

AFTER ABOUT A MONTH the vineyard was clearly visible. Small plants were everywhere, a breathtaking sight to those of us who had been planning for a year! They were growing! And there didn't seem to be too many gaps. Two months later they were small bushes. It seemed to us as though there were miles of them. We wondered what it would seem like when we got the whole thing in!

THAT YEAR THE stakes and the wires, even the irrigation went in after the planting. Subsequent years we surveyed, gridded and started with the stakes. There was the cable, the level and the eye-balling. Armando had been with us a year now and the crews were his. He was in total charge and they had become very good indeed.

The men are shown using the "diablo," the stake driver, familiar to those who drive posts but a little different in that it's shorter and has to be lifted higher. It weighs about thirty-five pounds. It is simply lifted and dropped. The effort is in going up, not pounding; thus it is a bearable operation. Even so, those who use it are paid extra. The men must use their eyes for the stakes' relationship to each other at every angle. There may be easier ways, but in any case those lovely satisfying lines don't come without attention.

Finally there was the installation of the end posts, the last touch that makes planted grapes look like a vineyard. A professional company was seen doing this nearby. We gratefully contacted them, not having known such a service existed. "Next?" we asked. "Fine," they said, and they came to see our arrangements, our stakes and what we might want. Their machinery is a modification of a post driver used by the highway people. In the case of a vineyard, it lays the post at a sixteen-degree angle and pounds it with a phenomenal ratatatat that's incredible to watch. Now and then the ground is so hard it is necessary to help with an auger. Following them is a tractor and trailer carrying wheels of wire strung through pulleys that allows a man on each side to grasp and staple it to the stake with an air gun, thus doing two rows at a time as well as two rows of wire at a time.

U.S.
ORDER
PATROL

FORD

THE LAST PLANTINGS shown here are about six weeks old, done as I've mentioned, everything before the planting.

After the first year of irrigation we decided we could put it in ourselves. Here are the hands that put in most of it. Warren Wellman, who's had something to do with every improvement that's been made on this place. Bill Bogue was our source of equipment as well as further information. We made many improvements, finally putting in a small pump at the bottom of the hill. The electric run was short, as there was an existing box we blessedly were able to hook up.

We had found that a bog continued to exist three months after the last rains in two areas. One we went into with French drains, taking up the soil in three long fingers ranging up the hill toward the base of the saddle. Perforated pipe was put in the bottom of it, tar paper over that, some six inches of gravel and finally the soil. The north-facing side we decided against fighting. We dug out more soil, put it in a big pile and made a dam. Now we have a pond there. (It was not as simple as that. It was complex and handsomely engineered by my old friend Tom Johnson.) It's now burgeoning with stolen fish, reeds and native plantings.

Finally we were able to take a look at the third leaf. Again here are Michael and Bret discussing the next decisions.

WE NOW HAVE forty acres planted. The biggest amounts are in merlot and sauvignon blanc. Our intention is to make a fine red wine and a fine white dinner wine. We will use small but adequate amounts of cabernet franc and cabernet sauvignon to spice our red and semillon to spice the sauvignon blanc. They are all in now and we are waiting for the ripe grapes of mature vines. Time and experiment will tell and we are very excited.

THE SANTA YNEZ VALLEY Viticultural Area is now established and well known for its fine wines. We expect some day soon we will be part of that venture.